THE READ AND WRITE SERIES: Book 4

# My Dirty Dog: My Informative Essay

By Darcy Pattison

Illustrated by Ewa O'Neill

My Dirty Dog: My Informative Essay
by Darcy Pattison, illustrated by Ewa O'Neill


Published in the United States.
For permissions contact:
Mims House
1309 Broadway
Little Rock, AR 72202, USA
MimsHouse.com

Publisher's Cataloging-in-Publication data

Names: Pattison, Darcy, author. | O'Neill, Ewa, illustrator.
Title: My Dirty dog : my informative essay / by Darcy Pattison ; illustrated by Ewa O'Neill.
Series: The Read and Write Series
Description: Little Rock, AR: Mims House, 2018.
Identifiers: ISBN 978-1-62944-092-7 (Hardcover) | 978-1-62944-091-0 (pbk.) | 978-1-62944-090-3 (ebook) | LCCN 2017919585
Summary: Cousins Dennis and Mellie spend the day at the local K-9 Karnival, write an essay about the day, then it's time for Grandma's birthday party!
Subjects: LCSH Family--Juvenile fiction. | Cousins--Juvenile fiction. | Dogs--Juvenile fiction. | Writing--Juvenile fiction. | Grandparents--Juvenile fiction. | BISAC JUVENILE FICTION / Family / General | JUVENILE FICTION / Animals / Dogs
Classification: LCC PZ7.P27816 My 2018 | [E]--dc23

**My teacher,
Mr. Eagle, said,**

"You must write an informative essay for homework. Bring it tomorrow. Remember, you need facts to write this essay."

I wanted to write about my dog, but I needed to narrow the topic. That means I can't write everything about my dog even though I love her so much. I had to think of just one small thing to write about.

My cousin Dennis was visiting for Grandma's birthday. After school, we all went to the K-9 Karnival at the city park. Dogs are also called "canines." In the Latin language, that means "dog." It's a joke to write it K-9.

# face painting

**The K-9 Karnival had many things to do. First, we stopped at the face-painting booth. I wanted a Maltese on my check because I own a Maltese dog named Lois Lane.**

**While she painted, the artist told me about my dog breed.**

"Your dog is an ancient dog from the Isle of Malta, which lies in the Mediterranean Sea. They have been bred for over 2800 years.

Some other names for the Maltese are Maltese Lion Dog, Cokie, Bichon, the Spaniel Gentle, and The Comforter."

**Dennis got a face paint of a Bernese Mountain Dog because of his dog named Clark Kent.**

**The artist said:**

"Yours is a Gentle Giant that came from the Swiss Alps, which are the mountains in Europe. It was bred as a farm dog to work with cattle or to work as a guard dog. Sometimes, they even pulled small carts. They call him a tri-color because he has three colors: black, brown and white."

Next, it was time for the Pooch Parade. Lois Lane and I wore matching clothes. We lined up with the toy group, or the small dogs. Dennis and Clark Kent marched with the working dogs. Dogs were dressed up crazy for that parade.

We were hungry, so Dennis and I ate hot dogs and drank milk. Clark and Lois ate dog food and drank water. After that, our dogs played in the dog park. But it had rained the day before so Clark Kent and Lois Lane were filthy.

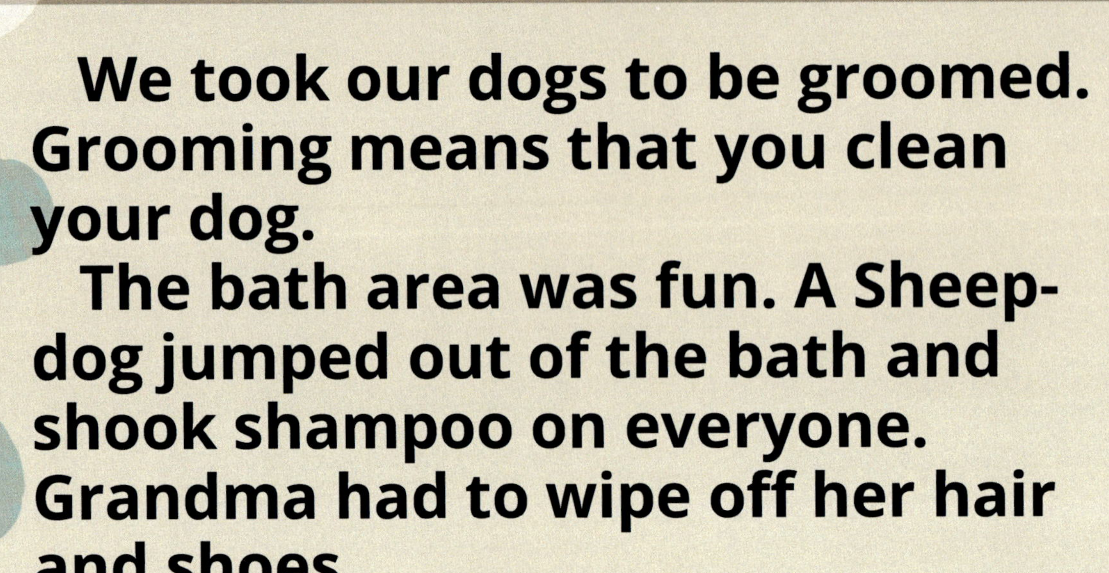

We took our dogs to be groomed. Grooming means that you clean your dog.

The bath area was fun. A Sheep-dog jumped out of the bath and shook shampoo on everyone. Grandma had to wipe off her hair and shoes.

When it was our turn, we got all the supplies ready: aprons, towels, brushes, shampoos, conditioners, and dryers. Then our dogs got clean while Dennis and I got wet.

K9
Clark Kent

Quickly, before they could get dirty again, we had K-9 photos made. We also got their paw prints made.

When we got home, I wrote my informative essay. When I thought about the K-9 Karnival, there was too much to write. I narrowed the topic to "how to groom a dog."

Here's my essay:

How to Give a Dog a Bath

If you have a dirty dog, it's time for a dog bath.

First, put on an apron because you want to stay dry. Then, gather supplies. To shampoo a dog, you need shampoo, conditioner, and lots of water. Wet dogs need towels, brushes, and hair dryers.

Start with one dirty dog. Carefully, get the dog wet.

Then, shampoo the dog. Be sure you don't get the shampoo in the dog's eyes because that will sting.

Next, rinse the soap off the dog. Now it's time to put hair conditioner on the dog and rub it in. Finish by washing off the extra conditioner.

Use towels to dry your dog's hair. Be gentle as you comb out the hair. Now, use a hair dryer to finish drying your dog's hair.

Lois Lane got a pretty bow in her hair. But your dog might not like that.

Now, your dog's hair is clean, but the veterinarian said, "Grooming means more than just a bath."

Use a dog toothbrush and toothpaste to brush its teeth. Look at your dog's toenails. If they are very long, use dog toenail clippers and cut them shorter.

Finally, give your clean dog a big hug.

**Hurrah!**
**My homework was done!**

**Dennis had to write an informative essay, too. Here's his essay.**

Dog Groups

Around the world there are hundreds of dog breeds. A dog breed is a group of dogs that are alike. The dog breeds can be grouped with other breeds that are alike in some ways. There are seven major groups of dogs.

Terrier Group

The terrier group has dogs that were bred to catch animals like rats. Terriers can follow an animal down a tunnel and catch it. They often have wiry hair.

Toy Group

The toy group has the smallest dogs. These dogs don't hunt. They are great as pets because they don't need much space in a house. For example, my cousin, Mellie, has a Maltese dog named Lois Lane. It's small and keeps my grandmother company all day.

Working Group

Working dogs were bred to do jobs. They might guard your house, pull a sled, or rescue someone in the water. For example, my dog, Clark Kent, is a Bernese Mountain Dog.

Sporting Group

The sporting group includes dogs who hunt in woods or water. If you hunt for birds, you might want a sporting dog. It will help you find the birds. Also, it will find the bird and bring it back to you.

Hound Group

Hounds are dogs used to hunt, too. Hounds can follow an animal by smell, by sight, or by both smell and sight. When they get close to an animal, some hounds bay loudly.

Herding Group

Herding dogs can control how other animals move. These dogs are like a living fence. Herding dogs can herd cows, sheep, ducks, or other animals.

These are the seven dog groups. There are many dog breeds, but they are divided into these groups. If you know these groups of dogs, it is easier to decide on a good breed for your family.

**Hurrah! Dennis's homework was done! It was time to party!**

We went to the kitchen, and there was Clark Kent and Lois Lane eating Grandma's birthday cake. Two dirty dogs!

We had to give them baths. Again.

And Mom had to go to the store to buy a birthday cake for Grandma.

After cake and candles, we played with our dogs.

Finally, Lois Lane yawned. Her mouth stretched wide, and it sounded like she said, "Aww."

I rubbed Clark Kent's belly, and he snorted.

Dennis snorted, too.

I snorted back.

# WRITING INFORMATIVE ESSAYS

## An informative essay gives information.

**This type essay might be a how-to essay with a sequence of instructions, or an essay that includes facts, definitions and details.**

**Choosing a topic.** Here are the guidelines for choosing a topic to write about. For a how-to essay, choose something about ten steps or less. More than that can be too complicated. Informative essays should be limited to one short subject. In the story, Mellie decided to only write about washing dogs. She didn't explain how to paint animal faces, or how to march in a parade. If Dennis tried to write about all the dog breeds, it would be a book instead of an essay. A good essay, however, might give information on one dog breed.

**Writing stronger details.** Informative essays include facts, definitions and details. Facts are any pieces of information about your topic. They can include size, color, shape, age, origin, or statistics. For difficult or unfamiliar words, an essayist should stop to define them. For example, Dennis defines a dog breed. Details are additional pieces of information. Often these break facts into smaller pieces. In Dennis's essay, he first give the fact that herding dogs control how other animals move. Next, he adds more details by giving example of animals that a dog might herd. "Herding dogs can herd cows, sheep, ducks, or other animals."

**Connecting words and phrases.** Good essays use connecting words to show relationships among ideas and facts. Use these words: also, in addition, another, and, or, but , for example, because, however, etc.

**Introduction and conclusion.** Establish the topic of the essay in the introduction. Provide a sense of closure at the end. Here are some ideas for an interesting introduction: cite a statistic, make a startling statement, ask a question, start with an interesting detail or sensory detail, describe a setting or character, give an example, reveal a secret, make a joke, or use a quotation. Interesting conclusions can include all the same things, as long as it gives the sense of closure.

Grandma?
She just snored.

67208782R00020

Made in the USA
Columbia, SC
25 July 2019